Samuel Gamble Bayne

The pith of astronomy

the latest facts and figures as developed by the giant telescopes

Samuel Gamble Bayne

The pith of astronomy
the latest facts and figures as developed by the giant telescopes

ISBN/EAN: 9783744737371

Printed in Europe, USA, Canada, Australia, Japan

Cover: Foto ©berggeist007 / pixelio.de

More available books at **www.hansebooks.com**

THE PITH OF ASTRONOMY

[*WITHOUT MATHEMATICS*]

THE LATEST FACTS AND FIGURES AS DEVELOPED
BY THE GIANT TELESCOPES

BY

SAMUEL G. BAYNE

WITH ILLUSTRATIONS

NEW YORK
HARPER & BROTHERS PUBLISHERS
1896

CONTENTS

ILLUSTRATIONS

INTRODUCTION

THE large modern telescope, celestial photography, and improved astronomical instruments have opened the field of astronomy to such an extent that the ideas, statements, and figures of a few years ago are no longer authentic. Happily for the credit of astronomers, the wonders of the skies have been underestimated, and those who thought that statements that had been previously made regarding celestial wonders were almost beyond credence will be pleased to find that they were not only true, but that in reality not more than half the truth had been told.

The writer has had the temerity to compile this little book in a simple and concise form for the use of those who know but little or nothing of astronomy, with the hope that it may lead them further to investigate this most delightful science. No mention has been made of signs, Greek letters connected with the naming of stars, or the mathematical features usually given in larger works on this subject, as they would only tend to confuse those who are seeking for elementary knowledge and to learn at least something of the wonders that surround us. When a lay reader has finished a large book on astronomy very little of the immense array of facts and figures can be retained in his memory. In this condensed form it is to be hoped he can remember much that will be interesting

and useful, should he go no further.
It is also intended to furnish a ready
reference for those who desire to re-
fresh their memories on the points they
have known but now forget, and to
give that information in corrected form,
from the most recent observations and
calculations, without the loss of time
incurred in searching larger works for
simple information. For those who
wish to recognize the constellations
and celebrated fixed stars at sight—and
what pleasure can be greater than rec-
ognizing them, as we do old friends,
when they make their annual reap-
pearance in the sky, if it be not that of
pointing them out to acquaintances,
who are usually eager to learn their
identity and something of their history?—William Peck's *Constellations,
and How to Find Them,* is perhaps the

best and easiest guide with which to find the stars; it contains but a few pages of reading - matter and twelve maps, which clearly show where the stars are on any night in the year.

The planets wander, as their name indicates, or rather they are changing their position to us at all times, and therefore cannot be located on a map, which if published would be incorrect in a month after its issue. However, the study of astronomy has of late years become so popular that most of the important newspapers employ astronomers to write articles on the existing condition of the heavens, and those articles naturally describe the location of the planets then in sight, so that any intelligent inquirer may find them with comparative ease. It may further be stated that a powerful opera-glass will show many wonders that are lost to the naked eye, and that the largest telescope that can be mounted on a tripod and conveniently moved out of the owner's residence for observing is a $3\frac{1}{2}$ - inch lens; a larger size than that needs a permanent foundation. It is an axiom in sight-seeing of all kinds that when fatigue commences instruction and pleasure end; and this will be verified in using a small telescope that needs the support of the hands and arms: it soon tires the observer. There is no satisfactory middle size between an opera - glass and a telescope mounted on a tripod.

The figures given hereafter represent averages (where there is a difference of opinion) taken from the highest authorities, such as Sir William Herschel, Professor Young, of Princeton, Sir Robert

Ball, Professor Langley, Camille Flammarion, Professor Lockyer, W. H. Warren, Garrett P. Serviss, and many others. The estimates, where it is practical, are given in round numbers, for the purpose of enabling the reader to remember them.

In beginning the order of chapters it will perhaps be best to commence at home, and our local solar system will be taken up first by a description of its members, starting with the centre, our sun, and running out to the frontier planet, Neptune, which is "one of the family," although it is more than two thousand seven hundred millions of miles from the earth.

THE PITH OF ASTRONOMY

THE SOLAR SYSTEM

"That very law which moulds a tear,
  And bids it trickle from its source,
That law preserves the earth a sphere,
  And guides the planets in their course."

THE natural division of the heavens
most interesting to the inhabitants of
the earth resolves itself into two great
parts—namely, the solar system, consisting of the sun with the eight planets
which revolve round it, and the great
suns or fixed stars which shine in space
at immense distances from the earth.

The solar group and its planets interest us most because we live within
its confines, and our earth is a part of

1

the system which enables us to observe closely and easily determine the dimensions, distances, composition, color, and weight of our neighbors; while the stars are so far from us that a large portion of our information regarding them is to some extent speculative.

The sun is the centre of the solar system. Eight planets revolve around it — viz., Mercury, Venus, Earth, Mars, Jupiter, Saturn, Uranus, and Neptune, in the order named—and are held in their orbits by its powerful attraction. The sun is a fixed star, of a low magnitude as compared with the giants of space. The word planet means a wanderer, and these bodies appear to wander through the skies, changing their positions daily; while a fixed star does not move—perceptibly, at least—in a hundred years.

The planets may be easily known

A RING THROWN FROM THE SUN FORMING A SEPARATE PLANET

from stars by the fact that, with the exception of Uranus and Neptune (revolving so far out in their orbits that they cannot be seen by the naked eye), they are larger to our sight than the stars, and shine with a steady flame, like a lamp, while the stars twinkle, as a bright point of light. Those who have telescopes may readily see the difference; a planet will show a distinct disk, while the most powerful instrument can only make the star point brighter and more brilliant. The cause of this is that the glass entirely fails to visibly magnify at such immense distances.

The moon is a sort of grandchild, viz., a satellite of a satellite—that is to say, it revolves round the earth while the earth is revolving round the sun. And, for that part, there is yet another step to take in this direction, for undoubtedly the sun itself, with its entire system of followers, is attracted towards a giant sun, compared with which our little one is very small indeed. Most of the other planets have moons revolving round them, which will be described in the proper place. It may be mentioned here that the four planets nearest the sun—namely, Mercury, Venus, the earth, and Mars—are the smallest and densest, and each turns on its axis in about the same time, while the four outside giant planets—Jupiter, Saturn, Uranus, and Neptune—rotate in about ten hours, and are the planets of the least density. The first four are known as the " interior planets," and the latter are called the " exterior planets."

An immense group of comets must be included in any description of the solar system, no matter how brief.

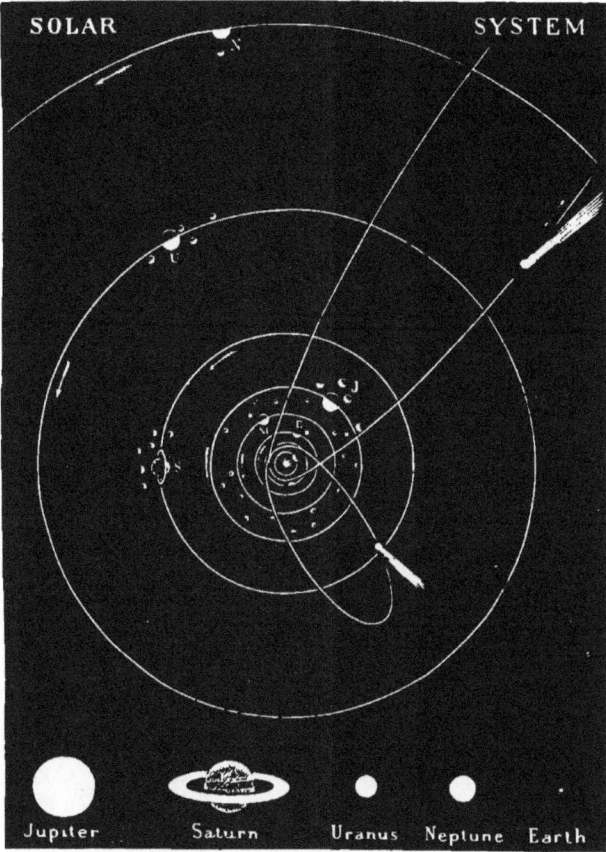

ORBITS AND COMPARATIVE SIZES OF THE PLANETS IN
THE SOLAR SYSTEM

These comets describe ellipses in their courses, and turn on the sun close to that orb, then run out into space, to return again in what is called their period—that is, the term of years that is consumed in completing their orbit. Some of them (notably Halley's comet) sail out far beyond Neptune, and as that would make a trip of six thousand millions of miles out and return, it is apparent that long intervals of time must take place between their appearances to us of the earth. In the vast distance between the orbits of Mars and Jupiter is found a great number of asteroids, as they are called. These also revolve round the sun, as do the larger planets, but they are so small by comparison that a more detailed description of them will be deferred till later.

Allied closely to these, but still smaller, are the meteoric stones, which are scattered through the skies, and rush round the sun in shoals, from the size of a walnut to that of a house. All of these members are the component parts of our system, running in size from the giant planet Jupiter, almost 1300 times larger than our earth, down to stones not as large as oranges; yet all these bodies have orbits, composition, and speed differing from each other, but each holding its proper place in the solar system which we call our own.

THE diameter of the sun is 866,000 miles.

Its mean distance from the earth is 93,000,000 miles.

Its volume, or bulk, is 1,300,000 times more than the earth.

Its mean density is one-fourth that of our earth, or about the consistency of a thick fluid.

It revolves on its axis in about 25 days.

Its mass—that is, the quantity of matter in it—is more than 800 times greater than all the planets combined.

The centre of gravity of the whole

solar system lies within the body of the sun when Jupiter and Saturn are not on one side of it.

The force of gravity at the sun's surface is 28 times greater than the gravity on the earth. A man weighing 217 pounds here would weigh over three tons on the sun, and his own weight would flatten and kill him.

Light travels at the rate of 186,000 miles per second, and reaches us from the sun in eight minutes. The light from the sun is 150 times as great as that of the lime cylinder of the calcium light, and it makes all other lights black, by comparison. With modern astronomical appliances the weight, size, and distance of the sun are now determined, and with the aid of the spectroscope scientists tell us the composition of the

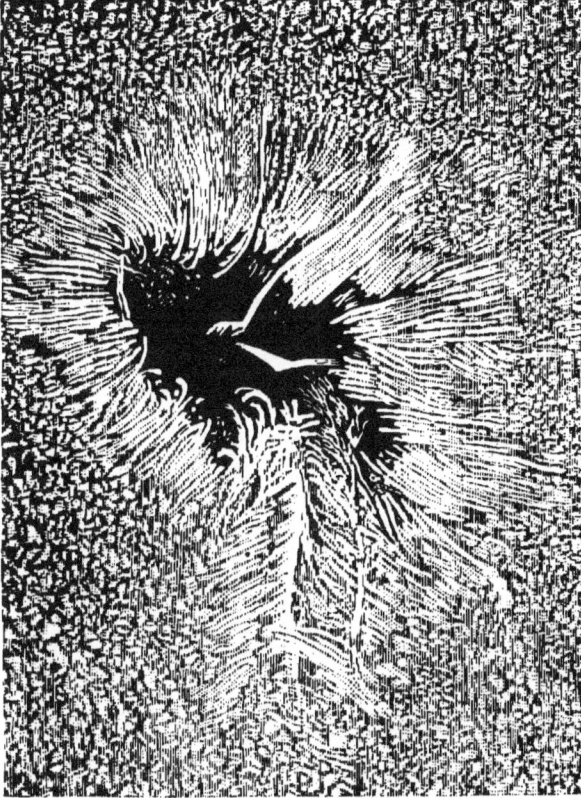

A GREAT SOLAR SPOT

(As seen by Langley.)

sun, as well as of many of the fixed
stars that surround us.

The sun is believed to be a mass of
intensely heated matter in the gaseous
state, consisting of both the permanent
gases, like hydrogen, and of metallic
vapors powerfully compressed by its
own gravity; this compression causes
it to contract, and the contraction,
which amounts to ten inches daily, pro-
duces the intense radiation that warms
and supplies the solar system with energy. If we take the sun's diameter
into our reckoning, we find that it may
require millions of years to perceptibly
affect this vast globe and reduce its life-
giving influence to a point where hu-
manity will freeze to death; but, after
all, it is only a question of time, as the
day will come when it will be reduced
to a dark cinder, and travel through

space as dead and cold as the moon is to-day.   It will then roam through the skies until perhaps it comes into collision with another body, when both will turn into a nebula of floating gas, thus forming the nucleus of a new world, which may be a home for new men and new things.   The sun, like all natural objects, must pass through the regular stages of birth, vigor, decay, and death.

The heat radiated by the sun defies all human conception.   It is a gigantic furnace, of such magnitude that comparisons between it and what we know of heat are futile.

A few of these may interest the reader.   If all the coal-fields on the earth were ignited and consumed in a fire, they would not supply the heat emitted by the sun for the tenth part of a second.

If the earth were to fall into the sun it would melt and evaporate on arriving there, like a flake of snow.

At the distance of 93,000,000 miles, were it not for the atmosphere that protects the earth from the sun's rays, they would melt a crust of ice enveloping the earth 100 feet in depth in a year, and would cause all the oceans to boil in a like period.

The sun's heat at its surface would also boil in an hour seven hundred thousand millions of cubic miles of water at the temperature of melting ice; and yet more than 99 per cent. of all this heat is wasted, not having touched any of the planets on its way. Estimating the total radiation of the sun at two thousand millions, the earth receives but one part of this. The noise of the terrific disturbance on the sun is of such power that it alone would kill a man were he placed within 5000 miles from its rim. The sun has a continuous evolution of heat equivalent to 10,000 horse-power on every square foot of its surface. A procession of icebergs sent against the sun would be melted at the rate of 300,000,000 cubic miles of solid ice per second.

If it were possible to have started an express-train for the sun in 1635, it would not be due there till now, and a single ticket for the trip would have cost $3,000,000. If a small community had started on the train, the seventh generation only would reach its destination.

The rays of heat from the sun on their way to the earth pass through a practical vacuum which has a temperature of 300 or 400 degrees below zero;

RELATIVE SIZES OF THE SUN AND PLANETS

they pass through this temperature and have apparently no effect until they meet some object, like the earth, capable of being warmed by them.

The sun has prodigious activity in its spots. These spots are sometimes 50,000 miles in diameter, and it is by observing them disappear and return to our sight by rotation that the time of the sun's rotation has been determined.

In 1858 a spot of over 107,000 miles in diameter was clearly seen. These spots are enormous vents for the tempests of flame that sweep out of and down into the sun.

An up rush and down rush at their sides have been measured at 20 miles a second; a side rush or whirl at 120 miles a second. These tempests rage from days to months at a time, and as they cease the advancing sides of the spots approach each other at the rate of 20,000 miles an hour; they strike together, and the rising spray of fire leaps thousands of miles into space; it falls again into the incandescent surge, and rolls over the Himalayas of fire as the sea over the pebbles on its beach. If ships were built as large as the whole earth, in such gigantic tempests they would be tossed like corks in an ocean storm.

The incandescent gases which are seen on the surface of the sun sometimes rise as high as 250,000 miles; they shoot out to these great distances often at the rate of several hundred miles per second.

The sun is our very life-blood; without it we could not live an instant. It directly supplies us with light, heat, and other forms of energy, and indirectly with food, clothing, and everything else we use. This provident worker has stored coal, petroleum, and natural gas for us in the past ages, thus giving us an inexhaustible reservoir of power and light. It furnishes us with wood, and lifts the waters to the hills, so that in their passage to the sea man may be enabled to harness them for his use in producing the necessaries of life.

It is the constant alternation of evaporation and condensation that keeps all the waters of the earth in a state of purity, making them fit for us to drink, and preventing the oceans from becoming stagnant. The salt breezes that sweep over the storm-tossed seas purify the air, and so on, *ad infinitum*.

We journey through a frigid space, but we live, as it were, in a conservatory in the midst of perpetual winter. We are roofed over by the air that treasures the heat, and floored beneath by a stratum both absorptive and retentive of heat—so that between the earth and air our vegetation and crops grow and ripen.

We owe these obligations and many more to the sun; it is little wonder, then, that men have bowed down and worshipped it in all ages.

THE planet Mercury is 36,000,000 miles from the sun.

Its diameter is 3000 miles.

Its year is 88 days.

It moves in its orbit at a speed of 29 miles per second.

Its day is supposed to be 24 hours and 5 minutes.

Mercury is the nearest planet to the sun. It can never be seen by the naked eye except in the west a short time after sunset, and in the east a little before sunrise. To see it, all the conditions must be favorable; it is said that the celebrated astronomer Copernicus never had the good fortune to see it.

2

It is usually lost in the glare of the sun, owing to its close proximity, and consequently the difficulty of observing it is great. Astronomers have not agreed on the length of its days or the density of its atmosphere.

Mercury is the densest member of the solar system.

Mercury has no moon.

When Mercury comes between the earth and sun, near the line where the plane of their orbits cut each other by reason of their inclination, the dark body of Mercury is seen on the bright surface of the sun. This is called a transit, and it occupies several hours in its completion, depending on the circumstances of the transit.

The next transit seen in this country will occur November 4, 1901.

VENUS

VENUS is 67,000,000 miles from the sun, and is the second planet from it.

Its diameter is 7700 miles.

Its year is 225 days.

Its orbital speed is 22 miles per second.

Its day is supposed to be a little over 23 hours.

It is sometimes called the evening star, and, as we see it, is the most magnificent planet in the solar system, exceeding in light and beauty the brightest stars; its light is so vivid that it casts a perceptible shadow, and we sometimes see it in full daylight.

The orbit of Venus is almost a perfect circle; the paths of all the other planets are more elliptical.

Venus has an atmosphere. It has no moon. Authorities do not agree on the length of the day on Venus, owing to the difficulty of observing the surface of the planet through its thick atmosphere.

An astronomical observation was made of Venus in Babylon in 685 B.C. It is written on a brick, which is now in the British Museum.

The heat on Venus is much greater than on our earth. Its water is supposed to be in the form of dry steam, the dense atmosphere causing the retention of its heat.

The dimensions of Venus make it a veritable twin of our earth, and in many other respects they resemble each other.

Land and water are believed to exist, and some of the appearances suggest the existence of mountains 27 miles high.

Venus is the ideal vision of the skies.

THE EARTH

"The globe terrestrial, with its slanting poles,
And all its pond'rous load, unwearied rolls."

THE earth is 93,000,000 miles from
the sun, and is the third planet from it.

The diameter of the earth is 8000
miles. (To be exact, 7926 miles through
the equator.)

It revolves round the sun in 365 days,
6 hours, 9 minutes, and 9 seconds.

Its circumference is a little less than
25,000 miles.

It turns on its axis in 23 hours, 56
minutes, and 4 seconds.

The earth is some 3,000,000 miles
nearer the sun in winter than it is in

summer; consequently, the sun looks larger to our eyes at that time, but the solar rays strike the earth in our hemisphere more obliquely in winter and do not produce so much heat.

The earth is as dense or heavy as it would be if composed entirely of metallic iron ore, which is five times heavier than water.

Geologists tell us that it is hundreds of millions of years since the earth was in the condition of a molten mass of matter, with its crust just commencing to form as it slowly cooled.

The earth is more rigid than glass, therefore probably no large proportion of its interior can be liquid, as many have supposed it to be. Its interior must be largely metallic.

It is estimated that the earth is inhabited by fifteen hundred millions of people. Its surface contains 187,000,000 square miles, three-quarters of which are water.

It may be estimated that something like four hundred thousand millions of men and women have lived since the advent of mankind.

This globe has eleven known motions. Among the most easily understood is its daily rotation on its axis, the passage over its orbit round the sun, and the motion towards the bright star Vega, towards which the entire solar system is flying.

Viewed from Venus and Mercury, the earth is the brightest star of the firmament, lit up by reflection from the sun.

The earth is flattened at the poles 27 miles, and this leads to the truthful but paradoxical statement that the Missis-

sippi River runs up hill, as its mouth is
three miles farther from the centre of
the earth than its source.

We fly through space at the rate of
more than 18 miles a second, seventy-
five times faster than a cannon-ball,
and pass over our orbit round the sun
in a year, the orbit containing 585,-
000,000 miles. We thus travel 1,500,-
000 miles daily through the skies, but
never over the same path, as we are
chained to the sun and follow its orbit.

We would be blown from the earth
like dust did we not share its momen-
tum, or if the envelope of air did not
proceed with us.

Were the earth suddenly arrested in
its flight, or if it came into collision
with another large body, the heat produced would be so tremendous that

the entire globe would be instantly
turned into gas and form a floating
nebula.

A soap-bubble in the wind could
hardly be more flexible and sensitive
to influence than the earth. If the
water became more dense, or if the
globe were to revolve faster, the oceans
would rush to the equator, burying the
tallest mountains and leaving the polar
regions bare. If the water should become lighter, or the earth rotate more
slowly, the poles would be submerged
and the bottom of the equatorial oceans
become an arid waste. No balance
turning to the 1000th of a grain is
more delicate than the poise of forces
on this globe. Laplace has given
us indisputable proof that the period
of the earth's axial rotation has not
changed the 100th part of a second

of time in 2000 years. Man cannot make a clock that will tell the hours for a single day with the exactness that this vast globe has done for all recorded time.

Sunrise greets a new 1000 miles at every hour; the glories of the sunset follow over an equal space some 12,000 miles behind. While the east and west are gorgeous with sunrise and sunset, the north and south are often more remarkable still with their aurora borealis and Magellanic clouds.

We in this latitude know but little of the glorious "Northern dawn." It prevails near the arctic circle, and there takes many forms — cloud-like, arched, straight. It streams like banners, waves like curtains in the wind; it is inconstant, and is either the cause or result of electric currents; it is often far above our atmosphere, sometimes as high as 600 miles.

The realm of this royal splendor is yet an unconquered world, waiting for its Alexander.

THE MOON

"As when the Moon, refulgent lamp of night,
O'er heaven's clear azure spreads her sacred
light,
Around her throne the vivid planets roll,
And stars unnumbered gild the glowing pole."

THE moon revolves round the earth at a mean distance of nearly 239,000 miles.

Its diameter is 2160 miles.

The volume or bulk of the earth is almost fifty times greater than that of the moon, and it would take 60,000,000 of moons to equal the sun.

The surface of the moon contains about the same number of square miles

that are found in North and South
America.

It completes its revolution round the
earth in 27 days, 7 hours, 43 minutes,
and 11 seconds, which is its sidereal month; the ordinary month, from
new moon to new moon again, is 29
days, 12 hours, 44 minutes, and 2 seconds.

It revolves on its axis exactly in a
sidereal month, and therefore always
presents almost the same face to the
earth; thus we never have seen the
far side of the moon, nor will we ever
see it. This circumstance causes the
moon to have the longest day (caused
by the sun's light) of any known celestial body—there are but twelve of them
in our year.

The moon travels round the earth in
its orbit at a speed of 37 miles a minute,

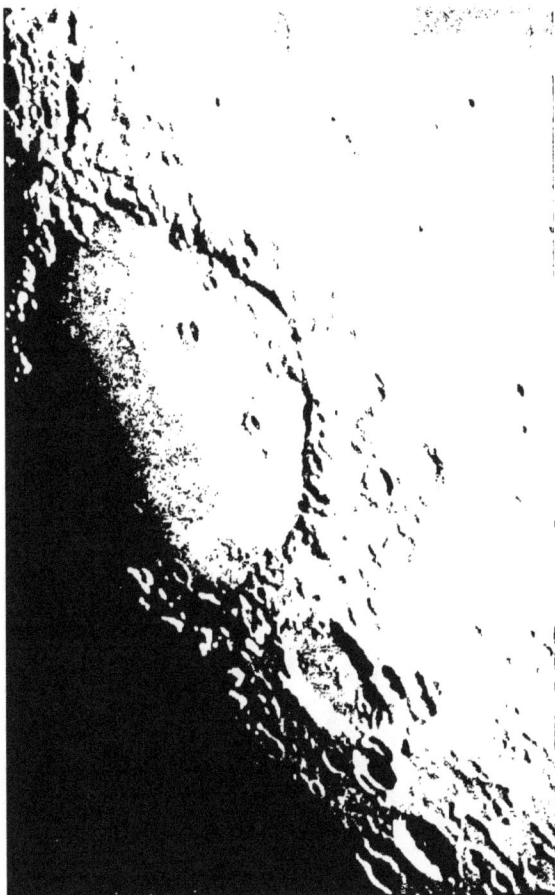

THE RIM OF THE MOON, SHOWING MARE CRISIUM

and its orbit contains about 1,500,000 miles.

The moon varies in size to the eye, as the distance from us varies to the extent of 25,000 miles in the course of a month.

When the moon is between us and the sun that side which faces us is not lit up by reflected light, and we do not see it; when it forms a right angle with the sun we see half of its face, and when we are between it and the sun we see it as the full moon.

A man weighing 155 pounds here would weigh but 26 pounds on the moon, and could, consequently, jump incredible distances.

The dimensions of the moon as compared with those of the earth are far greater than those of any other satellite in proportion to its primary.

The moon's day (caused by the sun's light) is almost thirty times as long as ours, and consists of fifteen days of daylight and fifteen days of darkness.

As it has no atmosphere to protect it from the sun's rays its heat in daylight is intense—strong enough to boil water—while at night the cold is frightful, being several hundred degrees below zero. Lord Rosse estimates the difference between day and night to be 500 degrees.

The moon is a dead cinder; if it ever had air and water, which it probably had, they are now absorbed in the porous lava that covers its surface.

In consequence of the small gravity at the moon, the absence of the expansive power of ice and the levelling influence of rains, precipices stand, mountains shoot up like needles, and cavities three miles deep remain unfilled; these conditions give the moon grand and savage scenery, such as cannot be found on the earth. It has twenty-eight mountains higher than Mount Blanc; ten of these are over 18,000 feet high, the two highest, Mounts Leibnitz and Dorfel, being almost 25,000 feet each.

These mountains have been measured with greater accuracy than any of our own, and in a general way the maps made of the moon are more reliable than those made of the earth. The extinct volcanic craters on the moon are enormous. The crater of Clavius has a diameter of 130 miles. By the aid of powerful telescopes 33,000 craters have been counted on the side of the moon which we see.

We are indebted to the moon for many things; but the greatest of these

3

is that it is principally owing to its attraction that we have the purifying motion of the seas known as tides. Without these daily currents the oceans would become stagnant and unhealthy to such an extent that we could not live on their shores.

MARS

THE planet Mars is 141,000,000 miles from the sun.

Its diameter is 4200 miles.

Its years contains 687 days.

Its mean distance from the earth is 48,000,000 miles.

The day on Mars is half an hour longer than ours, or about 24 hours and 37 minutes.

It has two moons.

It moves at the rate of 15 miles a second.

Mars is the fourth planet from the sun, and is called the red planet, from its well-known color.

The combination of its motion with ours causes it to pass behind us, or opposite to the sun, once in two years. For two months at this period it is best seen, and appears as a red lamp in the sky; at other times it looks small and unimportant.

Its density and size are less than ours; a man weighing 200 pounds here would weigh but 75 pounds on Mars.

The orbit of this planet is decidedly elliptical; it is 26,000,000 miles nearer the sun at the nearest part of its orbit than it is at the farthest, consequently the variation in heat from this cause alone is considerable.

In many ways Mars resembles our earth : it has atmosphere, seasons, land, water, storms, clouds, and mountains; it also snows and rains on Mars, as it does with us. Snow and ice cover

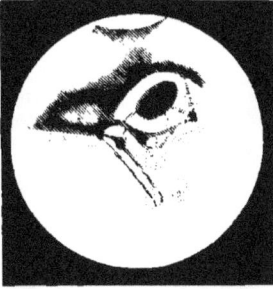

August 17, 1892

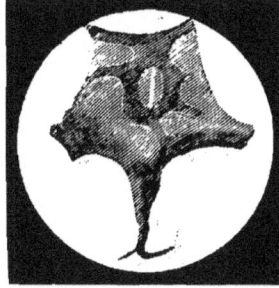

July 24, 1892

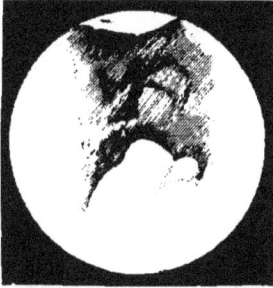

July 23, 1892

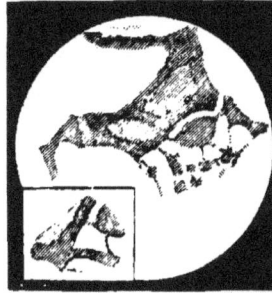

August 30, 1892

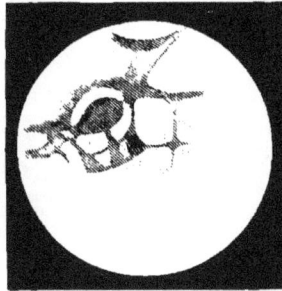

August 24, 1892

August 30, 1892

OBSERVATIONS OF MARS SHOWING ITS CHANGES

both its poles, and produce great white patches at those points, which are clearly seen through a large telescope; such an instrument also shows the markings on the land known as the canals. Fairly accurate maps have been made of Mars, showing its natural divisions of land and water to be about equal.

It has been suggested that the vegetation on Mars, for the most part at least, is yellow or orange, instead of green, as with us, thus giving the planet its color.

It is but 3700 miles from the surface of Mars to its nearest moon, and that satellite revolves round it in seven hours and a half, showing all the phases of our moon in a night; to an inhabitant of Mars it has the appearance of an enormous shooting-star slowly moving through the sky, and would take our breath away if we saw anything like it from our earth.

Percival Lowell, of Boston, has lately devoted his life and fortune to the observation of Mars. He has erected an extensive observatory at Flagstaff, Arizona, and has been using it for this purpose for two or three years. He is now providing a special telescope with a magnifying power of 2400 diameters, for the purpose of examining this planet.

Mr. Lowell's extended observations lead him to believe that Mars is inhabited by a highly civilized race of beings, who are now carrying on great engineering works, including the famous canals, which have been the subject of so much speculation.

JUPITER

"More yet remote from day's all - cheering
  source,
Vast Jupiter performs his constant course;
Five friendly moons, with borrowed lustre,
  rise,
Bestow their beams benign, and light his
  skies."

JUPITER is the ·fifth planet from the
sun, and revolves round it at a mean
distance of 483,000,000 miles.

Its year is almost twelve of ours—or
exactly 11 years, 10 months, and 17
days.

Its diameter is 88,000 miles.

Its volume is about 1300 times that
of the earth.

It is 390,000,000 miles from us when both Jupiter and the earth are on the same side of the sun.

The day on Jupiter is less than ten hours.

It moves over its orbit at the rate of eight miles a second.

Its light is so brilliant that it casts a shadow.

A man weighing 200 pounds here would, if carried to Jupiter, turn the scales at 500 pounds.

A web of cloth as long as from the earth to our moon would fall short of encircling this great planet. Jupiter is flattened at the poles and bulges at the equator, owing to the speed of its rotary motion, and if it rotated a little faster it could not keep itself together, but would burst, and be spread on the skies like a coat of paint.

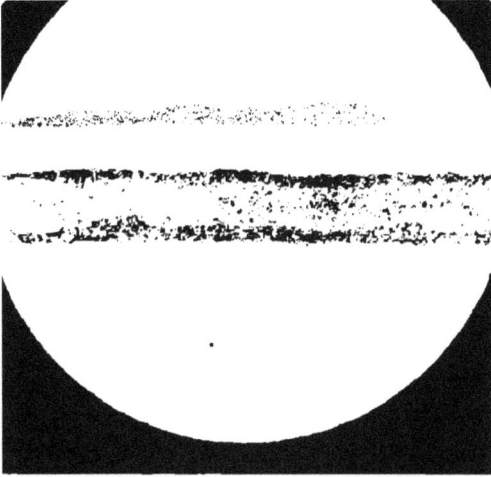

THE PLANET JUPITER AND ITS BELTS

Its days are so short, in consequence of the rapidity of its rotation, that its year contains 10,455 of them.

As its axis is vertical, there are no seasons such as we have, the most of its surface remaining in eternal spring.

The clouds in the thick atmosphere take the form of immense belts, on which spots appear, both of which can be plainly seen through a telescope; the atmosphere over the equator moves faster than the air north or south of it, producing the effect of a violent wind constantly blowing over its equatorial zone at a velocity of 250 miles an hour.

Jupiter has five moons; three of them are much larger than our moon, and one is larger than Mercury, having a diameter of 3600 miles. The nearest is 112,000 miles from the planet, and the most distant is 1,189,000 miles away; they travel over their orbits with varying speed, and, with their primary, are known as the Jovian system. It is very probable that these worlds are inhabited — more probable than that Jupiter has now any inhabitants—as they have atmospheres and some of the requirements for sustaining life. Jupiter seems to be a world in process of formation, cooling in preparation for the race that may at some future time occupy it. It has been said that this planet represents to-morrow, the earth to-day, and our moon yesterday.

The magnificence of the celestial spectacle presented by the Jovian system is beyond description, as seen by one standing on the nearest moon. Jupiter presents an immense luminous disk, more than 3000 times the size of our moon, while the sight is diversified by the other four worlds flying round in their orbits, and all comparatively close to the observer. These moons have a variety of color; two are blue, one is yellow, and one red. Jupiter spins like a top in the centre, the moons rush round it, and the whole procession sweeps through the skies at the rate of 500 miles a minute. Yet the disclosure of all this power, skill, and stability is but entering the vestibule of astronomy.

SATURN

"One moon to us reflects its cheerful light,
  There, eight attendants brighten up the night;
  Here, the blue firmament bedecked with stars,
  There, overhead, a lucid arch appears."

SATURN's mean distance from the sun is 883,000,000 miles.

It is the sixth planet from the sun.

Its diameter is 75,000 miles, and it is the largest planet excepting Jupiter.

It is 790,000,000 miles from the earth.

It revolves round the sun in 29 years and 5 months.

Its volume is 697 times that of the earth.

Its day is 10 hours, 14 minutes, and 24 seconds.

It has 25,000 days in its year.

Its orbital speed is 6 miles a second.

Eight moons revolve round it; no other solar planet has so many. In composition it is lighter than water, and it has a dense atmosphere.

The poles are flattened one-tenth of its diameter, which is a larger proportion than on any of the other planets.

On account of the velocity of its rotary motion, gravity varies greatly on the surface; the centrifugal force at the equator antagonizes gravitation to such an extent that while a man would weigh less there than he does here, at Saturn's pole he would weigh more than on the earth.

Our opportunities for seeing Saturn vary greatly. As the earth at one part of its orbit presents its south pole to the sun, then its equator, then the north pole, so does Saturn; and we, in the direction of the sun, see the south side of the rings inclined at an angle; next, the edge of the rings appears like a fine thread of light, and then the north side at a similar inclination. It occupies fifteen years in making all these changes.

Galileo, with the first telescope, discovered Saturn's ring in 1610. In 1612 the thin edge was turned towards the earth, and with his imperfect glass he could not see it; greatly discouraged, and believing he had been deceived, he exclaimed, " Is it possible some demon has mocked me?" He would never look at the planet again, and died without knowing that he had discovered the ring.

Saturn is surrounded by an enormous

flat, luminous ring, which is one of the greatest wonders of the heavens, and when seen through a telescope it compares favorably with any celestial sight.

This ring is about 175,000 miles in diameter, and the average estimate of its thickness is 75 miles.

The composition of the ring has caused much speculation. Laplace demonstrated that it cannot be solid and survive an hour; Peirce showed it could not be fluid and continue; other authorities proved the impossibility of its being composed of gas; and finally Maxwell showed it must be composed of clouds of satellites, some of them probably not larger than an orange, but all of them too small to be seen individually, and too near together for the spaces to be discerned. This theory is now accepted as correct.

The ring has three divisions; the innermost ring is dusky and transparent; in contact with it is the brightest ring, called ring B; then comes a gap, and then the outer ring, known as ring A. There are other divisions, but they are not permanent.

If the scenery of Jupiter is magnificent, that of Saturn is unique. Here we have a universe, a colossal system, a wreath of vast proportions turned to silver by the reflection from the sun, and eight moons revolving outside its limits—travelling like pearls strung on a silver thread. No one has ever seen Saturn come into the field of a large telescope for the first time without saluting the spectacle with exclamations of surprise and delight. Saturn is the wonder of the solar system.

URANUS

URANUS is the seventh planet from the sun, and comes fourth in the order of size. Its mean distance from the sun is seventeen hundred and seventy - eight millions of miles, and from the earth it is sixteen hundred and eighty - five millions of miles distant.

Owing to its great distance, astronomers have not been able to determine the length of its day, but it has been estimated at 11 hours.

Its diameter is 31,000 miles.

It takes 84 years to make its revolution round the sun.

4

Its volume is 69 times that of the earth.

It has an atmosphere.

It speeds over its orbit at the rate of 4 miles a second.

There is a great surprise in store for the observers of this planet. It has four moons, and they revolve round it from east to west, differing in this respect from the other planets, whose satellites revolve from west to east, and in about the plane of their equators, while the followers of Uranus revolve in a plane nearly perpendicular to that in which the planet moves—*i. e.*, this system rolls like a carriage-wheel, while all the others spin like roulette-wheels, the motion of the former being backward.

Uranus may be seen by the naked eye, under favorable circumstances, as a sea - green star of about the sixth magnitude. Up to the time when large telescopes were first used Uranus was mistaken for a fixed star by those who observed it.

On the night of March 13, 1781, Sir William Herschel saw through his glass that it had a disk and moved slowly through the heavens; this celebrated discovery deposed Saturn as the frontier planet, a position it had held from the beginning. As we will see in the succeeding chapter, this led to another discovery, and Uranus had, in turn, to give way to Neptune as the sentinel of the solar system.

Herschel called his discovery Georgium Sidus, in honor of his king and patron; the people called it Herschel, but astronomers finally decided on its present classical name as the proper one.

NEPTUNE

"Who there inhabit must have other powers,
  Juices, and veins, and sense, and life than ours;
  One moment's cold like theirs would pierce the
    bone,
  Freeze the heart's-blood, and turn us all to
    stone."

THIS planet is the eighth from the sun, and is third in mass and volume.

Its mean distance from the sun is two thousand eight hundred millions of miles, and from the earth two thousand seven hundred and seven millions.

Its diameter is 37,000 miles.

It revolves round the sun in 164 of our years.

It has an atmosphere.

Neptune is attended by one moon, which moves round it in about six days, at a distance of 260,000 miles; and it is remarkable that its motion is retrograde.

The telescopes at present in use do not show Neptune to us clearly enough to determine its diurnal motion, so the length of its day is unknown.

It has the longest orbit and the slowest motion of any planet, and although it travels over its annual path of seventeen thousand millions of miles at the rate of 200 miles a minute, it may still be called the tortoise of the skies.

Since its discovery it has not as yet completed a third of its first rotation round the sun, and as it will not finish its initial trip till the year 2010, no one at present alive will live to see that event.

Neptune is invisible to the naked eye; when seen through a telescope the light we see has travelled from the sun to it and returned to us by reflection, a double trip of five thousand five hundred and seven millions of miles, in something over eight hours. It would take an express train over 10,000 years to accomplish this task.

There is no object-lesson in the wonders of light that we can grasp so easily as this; it is the longest return of light that the heavens afford us.

Neptune is our frontier planet, and was discovered simultaneously by Adams, of England, and Leverrier, of France, in the autumn of 1846. Both believed it to exist from observing that its neighbor, Uranus, was retarded in its orbit by the attraction of some great unseen world. Both of these men gave their calculations to astronomers possessing large telescopes, directing them where to look for the great unknown— and both were successful.

This, the last great discovery in astronomy, was sensational in every way, and is a standing monument to the highest reasoning powers of the human mind. Had it been discovered by a mere survey of the heavens, one of man's greatest achievements would never have seen the light of day.

THE word comet is derived from the Latin word *coma*, meaning hair. Comets are celestial bodies which move round the sun in greatly elongated orbits, usually elliptical or parabolic.

A comet usually consists of a brilliant point surrounded by nebulous light, which extends backward in the form of a tail or train. The nucleus or head has an undetermined amount of solidity, but stars may be clearly seen through all comets.

Of the physical condition of comets little is at present known. Professor Young, of Princeton, states that a comet

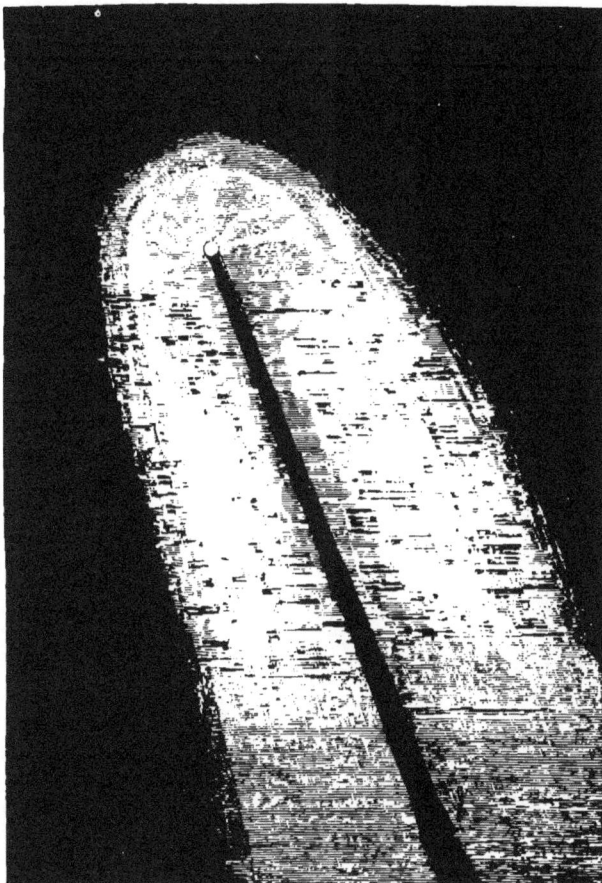

THE COMET OF 1881, AS SEEN THROUGH PROFESSOR
DRAPER'S TELESCOPE, JUNE 27

is nothing but a " sand-bank "; that is, it is a swarm of solid particles of unknown size and widely separate, say pinheads several hundred feet apart, each particle carrying with it an envelope of gas, largely hydrocarbon, in which gas-light is produced, either by electrical discharges between the particles or by some other light, the evolving action due to the sun's influence. This hypothesis derives its chief plausibility from the modern discovery of the close relationship between meteors and comets.

He also states that comets may hurt us in two ways, either by actually striking the earth or by falling into the sun, and thus producing such an increase of solar heat as to burn us up.

In regard to the possibility of a collision with a comet, Professor Pickering, of Harvard, says that it must be admitted that such an event is possible; if the earth lasts long enough such a thing is practically sure to happen, for there are several comets' orbits which pass nearer to the earth's orbit than the semidiameter of the comet's head, and at some time the earth and comet will certainly come together.   Such encounters will, however, be rare. If we accept the estimate of Babinet, they will occur once in 15,000,000 years, in the long run.

It is impossible to estimate, for want of sure knowledge of the state of aggregation of the matter composing a comet, when such a conflict will take place. If we accept the modern theory, and if this theory be true, everything depends on the size of the separate solid particles which form the main part of the comet's mass.   If they weigh tons, the bombardment would be very serious, but if, as seems more likely, the particles are smaller than pinheads, the result would be simply a grand meteoric shower.

Comets may be classed under two heads: those that return in their period—*i. e.*, in a stated number of years (the orbit of this variety is always in the form of an ellipse)—and those that travel in a parabola whose direction will cause them to run out into space and never return to the sun. These are named, respectively, periodic and parabolic comets.

Comets are further divided into those whose orbits lie within the solar system, and consequently have a short period, returning within a few years, and those whose path and direction carry them far beyond our system, returning after the lapse of centuries.

They are still further divided into those that can be seen by the naked eye and those that can only be seen by the aid of a telescope; the latter are known as telescopic comets, and the majority belong to this class.

It is estimated that 20,000 comets have passed within sight of the earth since the birth of Christ; of these 800 have been observed, but it is reasonable to suppose that there are thousands of millions of them moving in all directions in infinite space.

All comets that visit the solar system and turn on the sun are raised to incandescence from its heat when approaching and passing it; it therefore naturally follows that they become more brilliant when in its vicinity.

The comets of 1680 and 1843 were perhaps the most sensational that have

ever been seen by men; they were nearly
alike in splendor and dimensions. The
latter, flying at the inconceivable speed
of 340 miles a second, turned round the
sun from half-past nine till half - past
eleven on the morning of February 27,
1843. This is the greatest velocity that
has been definitely measured by astronomers; turning the celestial stake-post
gave them the desired opportunity. Its
tail was straight, and measured 198,000,-
000 miles, and it is one of the wonders of this observation that the tail
was always opposite to the sun, and
seemed not to be broken off or scattered on the skies when making such
an abrupt turn in two hours. Astronomers have calculated that it will return to us in the year 2219; we may
promise for our descendants that it will
have a memorable reception.

The comet of Donati, in 1858, was the most beautiful that has been observed; it had a brilliant head, and carried with it a curved tail measuring 55,000,000 miles. Its period is about 2000 years.

The comet of Pons has a period of 71 years; it appeared in 1812 and returned to us in 1883, and is due again in 1954.

Comets frequently have more than one tail. The comet of 1744 had six tails, which spread like a celestial fan over the sky.

Biela's comet was discovered by him in 1827, and it returned regularly in its period of six and one-half years. On its visit in 1846 it split in two defined comets, each having a head, coma, and tail. The twin comets returned "on time" in 1852, but have never appeared since; they have undoubtedly been lost, or captured by the attraction of some larger body, and will never again return to the sun.

Halley's comet is doubtless the most celebrated of all. Since the year 12 B. C. it has appeared on twenty-four occasions; its historical visits were in the year 1066, when William the Conqueror landed in England, and again in 1456, after the capture of Constantinople by the Turks.

On many of its visits the inhabitants of Europe were terror - stricken at its appearance.

It has a period of 76 years, and its orbit reaches out beyond the planet Neptune. Halley observed it on its visit in 1682, and, applying the principles of Sir Isaac Newton, he identified it as the comet of 1066 and 1456. Tracing it back to 12 B. C., and fixing its period at 76 years, he staked his professional reputation that it would return in 1758 (after his death), in the memorable lines: "Wherefore if it should return, according to my prediction, in the year 1758, impartial posterity will not refuse to acknowledge that it was discovered by an Englishman." It appeared on Christmas day, 1758, and Halley has since held a niche in the temple of fame.

We no longer regard a comet as a sign of impending calamity. We now look on them as interesting and beautiful visitors, which come to please and instruct us, but never to threaten or destroy.

ASTEROIDS OR PLANETOIDS

BETWEEN the orbits of Mars and Jupiter there is a space of nearly 400,000,-000 miles. Up to the year 1800 this belt was supposed to be vacant. On January 1, 1801, Piazzi, an Italian astronomer, discovered the minor planet Ceres. This was followed by an embarrassment of riches, and to-day more than 400 have been found.

Ceres is the largest. Professor Barnard gives it a diameter of 600 miles.

Vesta is the brightest, as it can be seen by the naked eye. Its diameter is about 250 miles.

Some of the smaller asteroids run

down to 20 miles in diameter and even lower, some authorities stating that there are shoals of them no larger than rocks.

These planets revolve round the sun (on an average) in less than five years.

STRUCTURE OF A TEXAN AEROLITE

SHOOTING-STARS

THE universe swarms with meteoric stones. These bodies, although very small, are of course not stationary, but revolve round the sun — that is, those that come within the domain of the solar system. They are governed by the same laws as the other bodies, and are a part of our system — a part of the unity of the universe. They have their region of travel, and the sun chains them and the giant Jupiter by the same power and influence.

When they come near enough to the earth they are "captured" by it, and rush into our atmosphere with such velocity as to produce a degree of heat sufficient to vaporize them and turn them into meteoric dust. During this process of burning they appear to us as shooting stars.

They usually become luminous about 75 miles from the surface of the earth, and are entirely consumed by friction before they descend to regions where the atmosphere is dense. In rare instances they are so large that they are not entirely consumed, but fall on the earth as meteoric stones in a heated condition. Specimens of them may be seen at most of our museums. These stones encounter the earth by day as well as by night, and simultaneously on all parts of the globe. Professor Newcomb has demonstrated that one hundred and fifty thousand millions of them annually fall on the earth. If they were

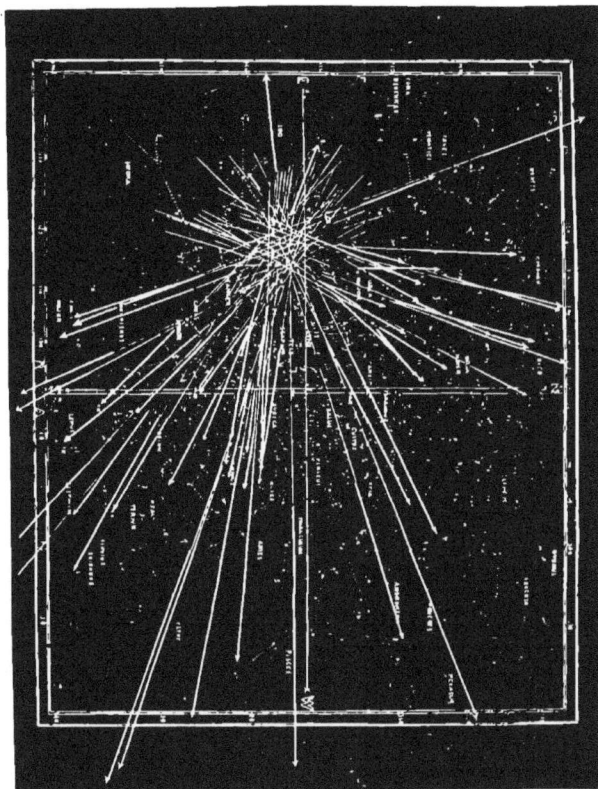

A SWARM OF METEORS MEETING THE EARTH

not melted through friction, we would be subjected to a continuous bombardment, destructive alike to our lives and our property.

Professor Peirce states that the heat which the earth receives directly from meteors is the same in amount which it receives from the sun by radiation, and that the sun receives five-sixths of its heat from meteors that fall on it, so that, after all, these little stones may play a more important part in our personal affairs than we give them credit for.

The dust caused by the destruction of these stones is often seen on the arctic snows, and is found at the bottom of the oceans, where it is subjected to but little agitation, owing to the comparative calm prevailing there. The writer has seen it taken from the bottom of the Indian Ocean by lowering a hollow socket partially filled with soft wax.

There are two thick streams of these stones that are annually encountered by the earth while its orbit is crossing theirs. The first meeting is on the night of August 10th; the second swarm is encountered on November 14th, at about three in the morning. The November belt is thin, and the earth runs through it in a few hours.

The August stream is more scattered, and we see these apparitions for some nights both before and after the 10th, which is the central date of their appearance in the skies.

There have been many remarkable star showers on these two dates—the most memorable in November, 1833; Olmsted, the astronomer, estimated that from the single point where these observations were made 240,000 stars fell during the shower, which lasted seven hours.

Stone showers and the fall of large stones have been more or less frequent in history.

Several thousands of stones fell in Hungary on June 9, 1866, with frightful noise; the largest fragment weighed 646 pounds.

The largest authentic aërolite is that found at Bendigo, Brazil, in 1816; it weighs five and a half tons; it was conveyed to Rio Janeiro in 1886.

On the evening of December 21, 1876, a meteor of unusual size and brilliancy passed over the states of Kansas, Missouri, Illinois, Indiana, and Ohio. It was first seen in the western part of Kansas, at an altitude of sixty miles. In crossing the State of Missouri it commenced to explode, and this breaking up continued while passing over Illinois, Indiana, and Ohio, till it consisted of a large flock of brilliant balls chasing each other across the sky, the number being variously estimated at from 50 to 100; it was accompanied by terrific explosions, and was seen along a path of not less than 1000 miles. The writer's recollection of the occurrence is that the stones were lost in Lake Erie.

While writing these lines the news comes through the Continental papers that one of the greatest meteorites ever seen recently shot across Spain, coming into sight over the Atlantic, and falling into the Mediterranean, or perhaps reaching the desert of Sahara, in Africa. On the morning of February 10, 1896, this body passed over Madrid with deafening explosions and a vivid glare of blinding light; it appeared as if enveloped in a bluish white cloud bordered with red; the city was shaken as with an earthquake, many windows were broken, and some light structures were levelled to the ground; the barometer fell, and afterwards rose rapidly. Nearly all of Spain was treated to a pyrotechnic display; incandescent fragments fell from the flaming meteor at Logroño and set the town on fire in two places.

There are many varieties of these bodies, differing in size, chemical composition, color, and origin: such as the ruins of vanished worlds scattered into space, the débris of comets, the product of volcanoes on stars and planets, which may include bodies projected from eruptions on our own earth in the violent vol-

canic disturbances that have undoubtedly taken place upon it in prehistoric
days; in other words, they represent the
accumulation of filings, fragments, and
dust from the celestial workshop in
the manufacture and disintegration of
worlds.

THE FIXED STARS

WE have finished our brief notice of the solar system and all that it contains which will interest the casual reader, and must therefore now cross the "Great Divide" to the infinite beyond.

What does crossing this divide mean? It means that we must leap the abyss from Neptune, our solar outpost, to the nearest body in space, the fixed star Alpha Centauri, some twenty-five thousand billions of miles distant; then another stride of nearly twenty-five thousand billions more, and we come to our second neighbor, the star 61 Cygni.

What has been said of our system will not more than represent the description of a few grains of sand as compared with the gigantic machinery that is in active operation beyond the confines and influence of the sun. The fixed stars will claim our first attention.

All the stars we see in the heavens are popularly called " fixed stars," to distinguish them from the wandering planets of the solar system. The name was given them by the ancients, when they were believed to be stationary. Modern science has shown that none of them are " fixed," but that every one of them is flying through the universe in some direction, with a velocity that is simply incredible to those who have not studied the subject.

The briefest reflection will satisfy any

intelligent mind that the laws of nature
will not permit any body to hang sus-
pended in space without motion; it
must move or fall in some direction,
and when it falls it moves; so, then, it
clearly follows that all is motion in the
heavens, and that nothing is "fixed"
or stable. All we either see or know
of fly, fall, roll, or rush through the void,
and yet, seen as a whole, all seems re-
pose. But it is not so. Each sun is
moving with a fearful velocity.

The equilibrium of the stars, like that
of our planets, is maintained by the exact balance of centrifugal force and the
attraction of gravitation; were it not
for the balance and harmony held by
these forces all would soon be turned
to chaos, and constant collision would
destroy the worlds in existence. The
velocity at which these bodies are mov-

ing is so great that, were they to meet, they and all they contain would be turned into vapor in less than a second of time.

In order to facilitate the indication of the size and brightness of a star, all the stars have been classed in the order of their magnitude. The word magnitude, however, is misleading, as it has no connection with the real size of a star, but simply indicates how they appear to our eyes. Thousands of stars that we can hardly perceive are undoubtedly giant suns; but, owing to their immense distance, they appear to us as mere specks on the firmament. This can be readily illustrated by looking at the moon and Jupiter in our little system; the moon is but a mere fraction of Jupiter in reality, but, owing to the difference in distance, Jupiter appears to us like a spark of light compared with the size of our satellite.

Astronomers have agreed on 19 stars of the first magnitude; of these, 6 are seen in the southern hemisphere, leaving us 13 to deal with. They are as follows, in the order of their apparent size and brilliancy :

		Of the constellation
Sirius,	. . . . .	of the Great Dog.
Arcturus,	. . . .	of the Herdsman.
Vega,	. . . .	of the Lyre.
Rigel,	. . . . . .	of Orion.
Capella,	. . . .	of the Wagoner.
Procyon,	. . . .	of the Little Dog.
Betelguese,	. . .	of Orion.
Aldebaran,	. . . .	of the Bull.
Antares,	. . . .	of the Scorpion.
Altair,	. . . . .	of the Eagle.
Spica,	. . . . .	of the Virgin.
Fomalhaut,	. . .	of Australis.
Regulus,	. . . . .	of the Lion.

The line between the last of the above list and the first stars of the second magnitude is very thin; many would, for instance, include the second-magnitude stars Castor and Pollux within the limit of the first. We count 59 stars of the second magnitude and 128 of the third; the stars in the succeeding magnitudes run into the thousands.

Of all the stars there are but 23 whose distance has been measured; the others are so far away that no angle or parallax can be found for them even by observing them from the opposite sides of the earth's orbit, which in itself is about 200,000,000 miles in diameter.

The last sun in this list that has been measured is the catalogue star, 1830 Groombridge, having a distance of four hundred and twenty-six thousand billions of miles. Of the distance of all

the hosts of heaven, with the above ex-
ceptions, we know nothing.

Alpha Centauri is the nearest sun;
it is twenty-five thousand billions of
miles from us. It would take an express train 73,000,000 years to reach it
from the earth.

The next nearest star is 61 Cygni;
it may be seen in the constellation of
the Swan on any clear night; it is
forty-nine thousand billions of miles
from this planet.

We know more than a million stars,
separately observed, catalogued, and
registered on celestial charts, but the
large modern telescope can now reveal
stars of the fifteenth magnitude, and
this brings hosts of new suns to our
knowledge, estimated to be at least
100,000,000 in number.

Celestial photography penetrates still

6

further and shows more, so that 20,000 stars are now known to exist for every one we see with the naked eye.

In the memory of man many stars have changed; some have faded from view, others have become brighter, while a few have changed their color, notably Algol, the variable star; it was formerly red, but is now white.

All the stars are moving in one direction or another. It takes Arcturus 800 years to move so small a distance as twice the apparent diameter of the moon, yet it is moving at the rate of sixteen hundred millions of miles yearly; its distance is so great it does not appear to move. Another example, Sirius, takes 1300 years, apparently, to move but a few inches in the sky, yet its minimum speed is about 2,000,000 miles every day.

It was not always safe to make such assertions, however, as Giordano Bruno was burned alive in Rome in A.D. 1600, by order of the Inquisition, for asserting that the earth was not standing still and was not the centre of the universe. Again in 1616 and 1633, Galileo, one of the greatest of astronomers, was imprisoned for the same cause, the Pope ordering that all books should be destroyed that asserted the motion of the earth.

Some stars are advancing to us, others receding at great velocity; but their size and the distance is so enormous that these factors count but little in their appearance to our eyes. Some of the larger stars flying towards us are: the Pole Star, at the rate of 46 miles a second; Vega, at 44; Arcturus, at 50; and Pollux, at 40 miles; while a few of those receding are: a Coronæ, at a speed of 48 miles a second; Castor, at 28; Capella, at 27; Regulus, at 23; and Sirius, the Dog Star, at 22 miles a second.

It will thus be seen that while the twins Castor and Pollux have stood side by side in the heavens—at least, to human eyes—for 4000 years, yet they are flying apart at the velocity of 68 miles a second, which in a day amounts to over 5,000,000 miles. This for 40 centuries, and still they are "the twins" to our eyes to-day.

The student of stars will in a short time begin to see that they have individual colors. Many of these are clearly marked, and the observer will at once notice their various tints. It is believed they take their colors from their chemical composition in a state of incandescence.

Aldebaran and Antares are red; Betelguese is orange; Sirius, Vega, and
Altair are bluish white; Arcturus,
Capella, and Pollux are yellow; while
others have tints of the ruby, the emer-
ald, the topaz, and the sapphire.

There are many variable stars—that
is, stars which grow bright and then
fade in a fixed period. Of this class
Mira and Algol are the most remarkable. Mira attains the size of a second - magnitude star and remains in
that condition 15 days; then it gradually fades and remains invisible for 5
months; afterwards it reappears and
increases slowly, to again become brilliant; the entire transformation occupies 331 days. Mira is sometimes called
"the wonderful."

One of the accepted explanations of
these changes is that it alternately

burns fiercely and smoulders within the time of its period.

Algol has a short, exact period of 2 days, 20 hours, and 48 minutes, in which it makes the change from its brightest to its faintest condition. It has kept time to these figures for 300 years. These changes are caused by the eclipse of this sun by an enormous dark satellite. The ancients named Algol "the demon."

The most astonishing change comes from the southern sky. In the constellation of Argo the star known as Eta, in 1843, disputed with Sirius the premiership of the skies. In 1856 it commenced to decrease, and gradually became smaller, till at length, in 1870, it left our sight. Seen at present with a telescope it is reviving, and may in the coming century regain its lost glory.

If this sun has satellites and they are inhabited, all their living beings must have perished from loss of heat.

In 1572 Tycho Brahé observed a new and bright star that suddenly appeared in the constellation of Cassiopeia. It was so bright it could be seen in daylight. It gradually faded from sight in 17 months, and has never been seen since.

The causes of these momentous changes in the great suns of space are largely a matter of speculation. There is little if any doubt that there exists a great number of extinct, burnt-out suns —enormous black balls that gravitate round other dark bodies — constituting dead systems. The dying throes of these monsters may have taken place in a celestial conflagration ending in their total darkness.

Another theory lately advanced by Professor Lockyer is that there is the closest possible connection between nebulæ and stars, and the first stage in the development of cosmical bodies is not a mass of hot gas, but a swarm of cold meteorites. Many bodies in space which look like stars are really centres of nebulæ—that is, of meteoric swarms— and stars with bright line spectra must be associated with nebulæ. Some of the heavenly bodies are increasing their temperatures; others, on the contrary, are decreasing. Double swarms, in any stages of condensation, may give rise to the phenomena of variability. New stars are produced by the clash of meteor swarms, and are closely related to nebulæ and bright line stars. Cosmical space is a meteoric plenum. The sun is one of those stars the temperature of which is rapidly decreasing, and many of its changing phenomena are due to the fall of meteoric matter upon the photosphere.

In point of speed, the most remarkable star in the universe is the seventh-magnitude catalogue star, 1830, Groombridge; it is invisible to the naked eye, but can be found with a glass, in the Great Bear. Its terrific speed is such that it has led astronomers to believe it is not propelled by any force we know of, but by some·power from another universe, and that it is simply a visitor passing through our skies. It is rushing on at a rate of 17,000,000 miles daily.

61 Cygni comes next to it in velocity. This star may be seen with the naked eye in the constellation of the Swan; its pace is over 100 miles a second.

The giants Arcturus, Vega, and Capella also move at a high rate of speed.

The light from the fixed stars is a
long time in reaching the earth; from
the very nearest it is about four years
in coming, and from the stars that are
so faint that we can hardly see them
it takes many thousands of years to
reach us.   Good authorities estimate it
in some cases as long as 100,000 years.
At the rate the stars are moving we,
then, never really see them, as they are
billions of miles away from the point
when the light we see left them.   It
also follows that succeeding generations
see stars that have become extinct for
thousands of years.

It is not the present state of the sky
which is visible, but its past history.

The earth, besides rotating on its axis
and revolving round the sun, reels like

a mighty gyroscope, but with so slow a
motion that it takes 26,000 years to
make one complete revolution of its
axis round an imaginary line perpendicular to the plane in which the earth
moves. Still further: as this axis of the
earth moves in its circuit round this
perpendicular line it points successively
to different parts of the heavens, and as
the point in the heavens to which the axis
is directed will not have any diurnal motion, all the stars will appear to revolve
round it, or round the star that may be
nearest to it; from which circumstance
it will be called the Pole Star.

Polaris, the present Pole Star, appears
"fixed" at the axis of the earth, but in
a few hundred years it will gradually
commence to wear away from it, and
in about the year 9000 our descendants
will elect Alpha Cygni as their guide,

while 13,000 years from now it will be the beautiful first-magnitude star Vega that will review the heavenly host.

Polaris has held the post of honor for over a thousand years, and was preceded in office by Thuban, of the Dragon, to see which in daylight the long tunnel in the Pyramid of Cheops was built.

The writer visited an observatory in China in 1874 said to be 4000 years old, in which were originally placed two bronze eyeholes on a slanting granite wall for the purpose of sighting Thuban, the Pole Star of that era. In 1874 the line of sight through these holes pointed to a void, all the stars having moved away from it.

THE CONSTELLATIONS

"Orion's beams! Orion's beams!
  His star-gemm'd belt, and shining blade,
His iles of light, his silv'ry streams,
  And gloomy gulfs of mystic shade."

FROM the very earliest ages the stars
have been watched with interest and
admiration, and their movements traced
and applied to various useful purposes.
In those days "the stars in their courses"
ruled the fate of men and nations.

For the purpose of identifying the
stars and finding out more about them,
the first watchers of the sky divided the
heavens into groups or constellations,
naturally naming each group after some

object to which they fancied it had a resemblance. As the first observers were chiefly herdsmen, we can readily conceive how some of the oldest constellations have been called after objects and animals with which the shepherds would be familiar in those early times.

Later in our history the Greeks set their mythological deities in the skies, and read the revolving pictures as a starry poem; they colonized the earth widely, but the heavens completely, and nightly over them marched the grand procession of their apotheosized divinities. The heavens signify much more to us, but we retain the old groupings and pictures for our convenience in finding individual stars. An acquaintance with the names, peculiarities, and movements of the stars at different seasons of the year is an unceasing source of

THE CONSTELLATIONS OF THE LION, THE HERDSMAN, AND THE GREAT BEAR

pleasure; one can never be alone if one is familiarly acquainted with the stars.

The constellation that is known to almost every one—young and old—is the Great Bear, popularly known as the Dipper, or Plough. The reasons for this are that it is a circumpolar constellation—consequently it is always in sight, revolving close to the Pole Star—and that it has a remarkable appearance, which all can recognize.

The two stars on the edge of the Dipper are near the Pole Star, and a line drawn through them points to it; they are, therefore, called the "pointers." This constellation contains no stars of first magnitude.

The group known as Cassiopeia, or Cassiopeia's Chair, is a companion to the Dipper, and is always opposite to it, as both swing round the pole.

The grandest constellation is the
giant Orion. It contains seven brilliant stars, two of them of the first
magnitude—viz., Betelguese and Rigel.
Three stars lie in an oblique line across
the middle of this group, and are known
as Orion's Belt. Flammarion calls this
group the California of the sky, because it not only contains the above
treasures, but in the middle of the belt· ⁹⁄₀
is found the wonderful nebula; viewed
through a powerful telescope there are
but few celestial sights that cope with
it in weird grandeur.

Canis Major, or the Great Dog, ad-
joins Orion, and is remarkable because
it contains Sirius, or the Dog Star—
the monarch of the skies and the greatest of them all in brilliancy and size.

Boötes, the Herdsman, contains the
giant flying sun Arcturus, which looks

so large that it may be mistaken for a planet. This star was recorded by Job and exploited by Homer. It is approaching us at a speed of 50 miles a second.

Ursa Minor, or the Little Bear, is interesting only because in it is placed Polaris, known also as the Pole Star or North Star. It is the guide of the sailor at sea, and has been used by the slave to point his way to freedom. It apparently never moves, or at least so little that its motion need not be discussed here.

Taurus, the Bull, claims attention because it contains a celebrated group, the Pleiades, mentioned by Job, and the theme of poets since writing began. The ordinary observer can see six stars in this group; many can make out seven; but Dawes, the keen-sighted

7

Englishman, counted thirteen under favorable circumstances. A recent photograph taken in Paris shows over 2000 in it. Aldebaran, the great red sun, is the eye of the bull, and may be easily located on account of its marked color.

The constellation of the Lyre is noted because of Vega, the most beautiful and one of the largest stars in the sky. It may be recognized by the formation of a small equilateral triangle with two minor stars. By the latest decisions of astronomers the solar system is flying towards this point. Vega can always be seen on a clear night, but is more brilliant when overhead in winter. This is true of all stars and planets; when they are on the horizon we have to look through so much more atmosphere that they become dim to our sight.

Leo, the Lion, may be known by the exact resemblance it now bears to a sickle. It contains Regulus, or the Lion's Heart, as well as Denebola. It is also a sign of the zodiac.

Aquila, the Eagle, contains the brilliant white star Altair, having small companions close to it on each side, the three making a straight line.

Cygnus, the Swan, may always be found in the Milky Way; it resembles a large cross or the skeleton of a kite more than it does a swan.

The other celebrated constellations containing first - magnitude stars are Auriga, the Wagoner, containing Capella, a brilliant yellow star ; the Little Dog, with Procyon within its limits ; the Scorpion, with the red star Antares ; the Virgin, containing the brilliant sun Spica as its feature. There are about a hundred other constellations, but they have little interest for those not making them a special study, and are comparatively modern, in most cases being named since the fourth century.

In addition to the constellations, the entire circumference of the sky has been divided into twelve parts, which have been named the twelve signs of the zodiac. This is the apparent path of the sun through the heavens. Their names and order may be easily committed to memory in the following rhyme:

"The Ram, the Bull, the Heavenly Twins,
  And next the Crab the Lion shines,
    The Virgin and the Scales,
  The Scorpion, Archer, and Sea Goat,
  The man that holds the Watering-pot,
    And Fish with glittering tails."

Some of these signs of the zodiac

are identical with the great constellations where they happen to lie in the
sun's course, notably the Bull, Lion,
Virgin, and Scorpion; the other eight
zodiacal signs are simply minor groups,
made memorable because they are the
mile-posts of the sun.

It was explained while describing the
fixed stars that in reality there are no
"fixed" stars; but they move so slowly
to our eyes that the ancients believed
them to be stationary. Their proper motion, however, causes them to move, and
therefore the dislocation of the heavens
is only a question of time. The day will
come by reason of this motion that the
neighboring constellations of Orion, the
Bull, and Canis Major will be merged
into one immense group. In 5000 years
this will be our largest and grandest
galaxy, and will require a new name.

Such a period seems long in human history, but it is but an hour in recording celestial time, when we consider that most astronomers agree in placing the age of our little modern planet at 20,-000,000 years.

Astronomers can go back in their calculations with as remarkable accuracy as they can predict the future. Fifty thousand years ago the Dipper had the form of a perfect cross, and in 500 centuries from now it will assume the shape of a steamer chair.

THE GREAT NEBULA OF ANDROMEDA

(Visible to the naked eye.)

NEBULÆ

A NEBULA is a luminous patch in the heavens, billions of miles beyond the limits of our solar system.

There are thousands of nebulæ, and they are of various composition, color, and size. Sir William Herschel observed and formed a catalogue numbering 2000; since his time many have been discovered by the aid of modern telescopes. He estimated that the light from the faintest would take 2,000,000 years to reach us.

Most of the nebulæ proper are composed of hydrogen and other gases strongly condensed; this is the nucleus

of worlds and planets in the process
of being condensed into a solid mass.
There are many worlds in such process
of evolution known to astronomers, and
they can be seen in their different stages
through a telescope. For example, one
in the constellation of Canes Venatici,
which shows a central condensation.
A second is found in Aquarius, and
shows a sphere in the process of throw-
ing off a ring; this ring may in time
condense and form a satellite. A third
may be seen in Pegasus, which is surrounded by rings of gaseous spirals.
Two of the most celebrated are found
in the constellations of Andromeda and
Orion; both can be seen on a clear night
by the naked eye, the latter surrounding the middle star of the three stars in
Orion's belt. This is the celebrated tra-
pezium nebula—a field of floating, glow-

THE GREAT NEBULA ABOUT THE MULTIPLE STAR ORIONIS,
IN THE CONSTELLATION OF ORION

ing gas, so large that our entire solar system would be lost in it. It is one of the great startling sights of the sky, and those who are privileged to see it through a large instrument can never forget it.

. In many cases these nebulæ are the graves of dead worlds and the cradles of new ones — immense masses of un-organized matter that may have been left floating in space — the wreckage from collisions of suns, now ready to revert back, in the process of time, to their original condition, thus sustaining the trite saying that nothing is lost in nature.

THE Galaxy, or Milky Way, is a luminous band of irregular form, consisting of a great circle entirely surrounding the heavens. It contains myriads of stars, so crowded together that their united light only reaches the unaided eye; this band of stars can be seen on any dark, clear night. If we could stand where the earth is and have it removed, we could see this splendid circle completely surrounding us; it is thus reasoned that we are a part of the Milky Way, and that our sun is near the centre of it.

The circumpolar constellations Cas-

THE CELEBRATED CRAB NEBULA NEAR TAURI

siopeia and the Swan are always to be found in the Milky Way, while Sirius, Capella, and Aquila may be seen on its edge when they are in sight.

The formation of the Milky Way assumes the general appearance of a silvery ribbon, but in places it is divided into two great branches, which afterwards reunite. Between these divisions are dark places comparatively devoid of stars; one of these, the Coal Sack, has become celebrated, and was so named by sailors because they could see no stars in this dark spot.

This glorious celestial path has been the theme of poets in all ages. Some of the best lines written about it are by Elizabeth Carter, from which the following is a selection:

"Throughout the Galaxy's extended line
Unnumbered orbs in gay confusion shine,

Where ev'ry star that gilds the gloom of night
With the faint trembling of a distant light
Perhaps illumes some system of its own
With the strong influence of a radiant sun."

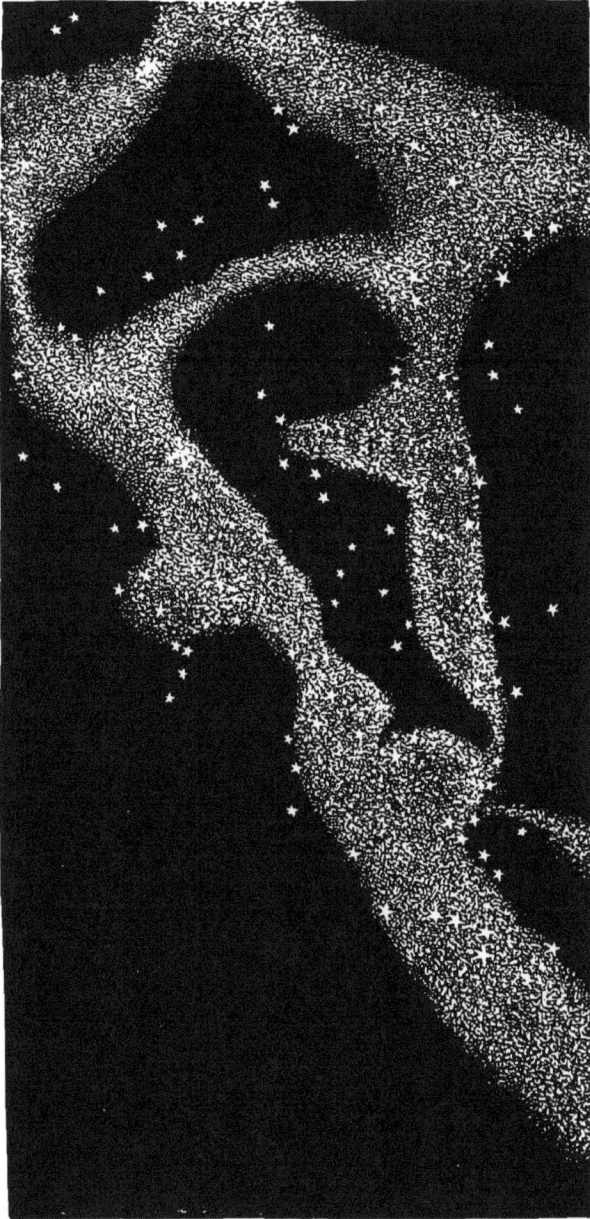

THE NEBULA IN THE MILKY WAY

DOUBLE AND MULTIPLE STARS

MANY stars that appear to the naked eye as a single object, when examined by a telescope or opera-glass are found to be composed of two, three, four, or even more stars. These are named double, triple, quadruple, and multiple stars. Some of these are in no way connected with each other save by the accident of perspective, while in many cases they compose a system, and revolve on one another in a fixed period; the periods may vary from a few years up to thousands.

Ten thousand double stars have been observed; the great majority of them are really double ; by the chance of perspective they are sometimes located almost behind each other, although really billions of miles apart.

There are 558 orbital systems known; 23 ternary systems exist, while 32 triples are made up of a binary system ; and an accidental optical companion, Kappa Pegasi, revolves round its partner in 11 years, which is the shortest period known.

Zeta Aquarii has the longest period definitely fixed ; it is 1624 years.

There are many systems whose periods exceed 5000 or 6000 years, but enough time has not elapsed since first observed to exactly fix their period.

The most celebrated double stars are Sirius and Castor, of the Twins. The former has a period of 53 years, and the latter has a companion that revolves round it in 1001 years.

The ternary system of Zeta, in the constellation of the Crab, is composed of three suns; the second revolves round the first in a period of 59 years, and the third round both stars in 600 years.

Double stars have almost always different colors, and frequently exhibit a brilliant variety of tints. Beta, of the Swan, contains two suns, one being golden yellow and the other sapphire. Antares and its companion are orange and green, respectively.

Procyon is so perturbed in its motion that it is known to have a large dark companion, whose attraction affects it, but the distance is so immense that no observer has yet been able to find the mysterious partner.

The celebrated star 61 Cygni is a telescopic double sun ; the constituent parts of it are forty-five times as far from each other as the earth is from the sun, yet it takes a powerful telescope to show any distance between these companions. No better illustration of the vast scale on which celestial mechanics are carried on can be found than by reflecting on this proposition.

These are but the bare, imperfect rudiments of astronomy, and represent but a taste of what is to come for those who wish to pursue this delightful science in its details. The author hopes that this little effort will not have been made in vain ; it certainly will not if it induces some of his readers to take up the subject in earnest where he has laid down his pen.

" What though no real voice or sound
  Amid their radiant orbs be found ?
In reason's ear they all rejoice,
  And utter forth a glorious voice ;
Forever singing as they shine,
  ' The hand that made us is divine.' "

NICHOLAS COPERNICUS. Born at Thorn, Prussia, A.D. 1473. To Copernicus must be given the first place in astronomy, for it was he who, in the face of all traditions, founded the Copernican system: placing the sun in the centre, with the planets revolving round it. Previous to 1543 all astronomers placed the earth in the centre of the universe, and believed that the stars and planets revolved round it.

GALILEO.—Born at Pisa, Italy, 1564. He discovered the properties of the pendulum in 1583, constructed a thermome-

YERKES TELESCOPE

(In the possession of the Chicago University. The total weight of this
instrument is 75 tons.)

ter in 1597, and invented and constructed the first telescope in 1609. With
these appliances he made many important astronomical discoveries.

His was the greatest opportunity
given to man in the field of exploration,
as the new glass placed him where no
one had stood before; but the invention
was his, and he used it to the fullest extent. He was imprisoned in Rome for
accepting the Copernican system.

SIR ISAAC NEWTON.—Born in England
in 1642. Newton was the greatest
mathematical astronomer, and was a
veritable wizard with figures, distanc-
ing all men who had lived before him
or who have appeared since.

The story of the fall of the apple
was first told by Voltaire, who obtained
it from Newton's niece. Laying down

the laws of universal gravitation was his principal work.

Newton was a philosopher as well as a great astronomer, as the treatment of his favorite dog will show. The documents completing a great work occupying forty years were spread on his library table; his dog upset a burning lamp and destroyed them. On his return to the room Newton affectionately patted the dog and exclaimed, "Diamond, Diamond, thou little knowest the damage thou hast done!"

SIR WILLIAM HERSCHEL. — Born in 1738. He discovered the planet Uranus and many other celestial bodies. With his own hands he constructed the first great telescope. It has been said of him that in nearly every branch of modern physical astronomy he was the pioneer. He was the virtual founder of sidereal science. As an explorer of the heavens he had but one rival—his son.

LIGHT is that form of luminous energy which comes to the eye in succeeding waves or vibrations at the rate of four hundred and fifty trillions per second. It travels at the rate of 186,000 miles in a second, and this uniform speed has had much to do in settling some of the greatest astronomical problems. Without its aid we would know absolutely nothing of astronomy. The division of light from the sun by a prism results in seven colors, in the following order : violet, indigo, blue, green, yellow, orange, and red.

This is known as the solar spectrum. The most extraordinary phenomenon connected with light and at present en-

tirely unexplained is that every star in
the sky is a centre of constant undula-
tions in all directions, which thus per-
petually cross each other through space
without being confused or mingled in
any way.

THE SPECTROSCOPE.—This wonderful
instrument makes known to us the composition of celestial bodies. With a
system of prisms it divides the light
from them into lines on a band or ribbon. The order and position of these
lines denote the chemical composition
of the body under examination, so that
we can determine exactly all the sub-
stances that compose it and their per-
centages. It is, in fact, the autograph
of the substance, written with lines in
colors. The astronomer by its aid can
as easily tell what the sun or Sirius are

composed of as a chemist can analyze the composition of gunpowder.

No Light without Dust.—The majority do not know that the sky is blue on account of thousands of millions of atoms of dust floating in the atmosphere. Were it not for dust we would lack light on earth and the heavens would be an inky black.

Suppose a room absolutely dark save a hole through one of the shutters. A ray of light will dart through the small opening, and one can observe tiny particles of dust dancing in that bright beam of light. As a matter of fact, it is not "the light" we see, but simply a reflection caused by these motes of dust.

As it is with this shaft of light in the darkened room, so it is on a large scale throughout the air. The many millions of particles of dust catch the light, reflecting it back and forth from one to another, so making the atmosphere luminous.

Were it not for dust the sky by day would appear black, as it does at night when there is no moon. The sun would appear as an immense glowing ball. The moon and the stars would be visible throughout the day. Everything would appear differently. Where the light touched, the eye would be dazzled by the brilliancy. The mellow softness of the shadows would become an intense black, and the outlines of objects harsh and angular.

The sunlight, which has been analyzed by the spectroscope, consists of all the colors of the rainbow, their total forming the white light. The white light going through a crystal prism is broken up into seven component parts, the so-called fundamental colors. These seven distinct colors of light are the result of the different lengths of ether waves. Blue is one of the shortest, yellow one of the longest waves. Thus the finest dust molecules floating highest in the atmosphere reflect only the blue light, imparting that tint to the heavens.

TYCHO BRAHÉ, the celebrated Danish astronomer, was born in 1546. Frederick II., of Denmark, noticing his remarkable talents, built an observatory for him on the island of Huen, and here it was, without the aid of a telescope and with the crudest instruments, he made observations that afterwards in the hands of Newton and Kepler were destined to settle the great problems of astronomy. It is asserted that he has never been surpassed as a practical astronomer, although he rejected the Copernican theory. He was eccentric in his conduct, and never made an observation in his observatory without putting on his court dress, alleging that if men dressed in honor of the king and court they should not be less observant of such duties in the presence of their Maker.

THE END

POPULAR ASTRONOMY

By SIMON NEWCOMB, LL.D., Superintendent American Nautical Almanac ; formerly Professor U. S. Naval Observatory. With One Hundred and Twelve Engravings, and Five Maps of the Stars. 8vo, Cloth, $2 50 ; School Edition, 12mo, Cloth, $1 30.

Its purpose is to enlighten that great mass of fairly educated people who have lost the astronomical knowledge that they once possessed. It states the latest methods of investigation, the latest discoveries, and the latest general development of this majestic and almost infinite science. Great thought and much space have been given to the historical points and philosophical aspects of the science. . . . In the treatment of weighty and abstruse scientific subjects, he never fails to bring them within the range of the average popular comprehension.—*Boston Post.*

The great reputation which the author of this work has merited and enjoys, both in this country and in Europe, is a sufficient guarantee of its excellence. . . . He has dwelt especially upon those topics which have just now a popular and philosophic interest, carefully employing such language and such simple explanations as will be intelligible without laborious study. Technical terms have as much as possible been avoided. Such as were employed of necessity, and many that occur elsewhere, have been fully explained in a copious glossary at the end of the book. With its abundant aid the reader cannot fail to derive both pleasure and entertainment from the study of what is the most ancient as well as the most elevating and inspiring of all the natural sciences. . . . Professor Newcomb, throughout his whole volume, preserves his well-known character as a writer who, in treating of scientific subjects, fully understands the art of bringing them within the range of popular comprehension. . . . . It is fully calculated to hold the attention of the general reader.—*N. Y. Times.*

PUBLISHED BY HARPER & BROTHERS, NEW YORK

☞ *The above work is for sale by all booksellers, or will be sent by the publishers, postage prepaid, on receipt of the price.*

RECREATIONS IN ASTRONOMY

With Directions for Practical Experiments and
Telescopic Work. By H. W. WARREN, D.D.
With Eighty-three Illustrations and Colored
Plates. 12mo, Cloth, $1 25.

Written not only to reveal some of the highest achievements of the human mind, but also to let the heavens de-
clare the glory of God. With sentiments of profound devo-
tion, and with the calmest belief that religion gains by
every advance of science, he invites the reader to scan the
heavens, and there find proofs strong as holy writ of the
truths of revealed faith. Dr. Warren writes like a scholar—
clearly, tersely, elegantly.—*Chicago Times.*

The style of the author is flowing and easy, so that even
his most scientific pages will make the reader pause and
catch the drift of the writing. The book will more gener-
ally interest readers that most books upon scientific subjects. It has an enthusiasm which is contagious. The
author has mastered well the art of bringing science into
the range of the common reader and making it both pleas-
ant and profitable.—*Chicago Inter-Ocean.*

The explanations of difficult matters are particularly lucid,
and for readers not technically instructed in astronomy
nothing could be better as a literary presentation of the attractive side of the science.—*N. Y. Evening Post.*

A very attractive book . . . treating the subject in so familiar a manner as to make the practical and useful information it contains delightful reading.—*Boston Commonwealth.*

PUBLISHED BY HARPER & BROTHERS, NEW YORK

☞ *The above work is for sale by all booksellers, or will be
sent by the publishers, postage prepaid, on receipt of the price.*